C.N. FARCAȘ

THE ENGINEERING OF THE SPIRIT

2020

Copyright © 2020 C.N. FARCAȘ

All rights reserved. No part of this publication may be reproduced, distributed, or transmitted in any form or by any means, including photocopying, recording, or other electronic or mechanical methods, without the prior written permission of the publisher, except in the case of brief quotations embodied in critical reviews and certain other noncommercial uses permitted by copyright law.

România, 2020

ISBN: 979-857-5-93624-4 (Black & white)

For information contact:

C.N. FARCAȘ
www.sacraf.com

TABLE OF CONTENTS

In search of behavior 6
A brief introduction about who is giving you all this information 9
What is the spirit? 13
THE SPARK 16
Understanding the spark 17
Balanced spark 21
THE WILL 25
Action 26
Inner voice 28
Meditation 29
Balanced spiritual state 30
Discipline 33
THE THREE ENERGIES 34
Everything starts with belief 35
Reactions 37
Situations & events 38
The 3 black dots of darkness 39
The manifestation of darkness 43
Chaos 47
Loss 48
Death 49
Two worlds 50
Interior tools 51
Colors 53
THE RED REACTION 54

THE YELLOW REACTION	56
THE BLUE REACTION	57
Overview	58
THE SPIRIT	62
DARKNESS & LIGHT	63
Structural representation of the spirit	64
The spirit along with darkness	65
Spirit, darkness, belief, and a few emotions	66
Superposition, fission, fusion	67
Mirror spirits	72
Connections	73
Breathing	74
THE THREE WARRIORS	76
There are 3 warriors inside us	77
The Red warrior	78
The Yellow warrior	80
The Blue warrior	82
The three voices and the Inner voice	84
THE WORLD BELONGS TO THE STRONG	86
MASTERING DARKNESS	87
Mastering chaos	89
Mastering loss	94
Mastering death	98
MASTERING LIGHT	102
Judgement	103
Two powerful methods to empower faith	106

Mastering the Red warrior 108
Creating darkness 112
The inevitable battle 115
Mastering the Violet warrior 117
THE WORLD BELONGS TO THE INTELLIGENT 120
Intelligence 121
Curiosity 122
Understanding 124
Imagination 125
Wisdom 126
THE WORLD BELONGS TO THE ADAPTOR 128
The three types of situations & events 129
THE APOGEE OF THE SPIRIT 131
PROBLEM SOLVER 133
BALANCE IS THE WAY 134
The purpose of energy 135
HOME 136
The gift from our parents 137
WE ARE NOT ALONE 138
UNITY 139

In search of behavior

There is something inside us that is constantly traveling throughout our whole nervous system. Throughout our body. Throughout our mind. **Throughout everything!** It has access to our muscles. It has access to our thoughts. It is permanently active and alive. It is like a **fiery water** and behaves as the electric current. It has already a name, but the passing of time and lack of research has made it fade throughout history.

The best analogy to describe this entity is by describing a **telephone**. We are similar to a cellular phone. The phone has its case. We have our **body**. The phone has its processor. We have our **mind**. The phone has its battery. We have our **subconscious**!

What lies beneath our body is the mind. The mind gathers information from our senses and process them forming consciousness! But there is something beneath consciousness. Something responsible for our life. Something known as **subconscious**.

We are not our height, our weight, the color of our hair, the color of our eyes, the color of our skin, or our genitalia. We are not our mind, our thoughts, or our way of thinking. The brain is the most powerful component of our body, but it is just a tool. A tool for us to use! We are our **subconscious**, which is our life source. In it lies our behavior, energy, and identity.

Inside us, we have a "fiery water". It moves with fluidity like water and burns with intensity like fire! It is our battery. It is our energy! It travels throughout our body giving us the ability to move. It travels throughout our central nervous system giving us the ability to think. With it, we can move, think, and do everything we want! We cannot see it. It is not an organ. Dissection is not an option. It behaves as an electrical current. It has changeable **amplitude**. It can create **pressure** as voltage does. The body acts as a **resistor**, allowing conduction and insulation to occur inside. When our energy rises, **heat** in the body can be felt. This energy is a **fiery water**, and it moves as a stream. Our body is its host. Our body is its <u>vessel</u>.

Our material world is composed of small, individual particles called atoms. In every atom there is something known as nucleus and it responsible for everything that has mass. The atom has something else as well. It has the **electron**! Electrons are responsible for generating electric energy.

Our fiery water behaves as the electric current. Its electrons flow inside us and form a stable structural energy. If we want to understand ourselves, we could proceed through two methods. Through theory, by gaining an understanding of electrons and their behavior. Or through practice, by understanding our behavior throughout exposure in physical activities such as sport.

This fiery water is not a simple energy. It is a **structural energy**! It is bounded by itself. It has an infinite time duration. The end of this battery is when the <u>case</u> of the "phone" or its <u>processor</u> erodes. And not vice versa.

A name has been given to our battery by our elder ancestors. The subconscious has already a name. It is called **SPIRIT**.

For too long we viewed it as a concept and not as a practical thing. That is because we **could not see it**. It was viewed as a myth or as something inexistent. As the electron was viewed. But when technology evolved, and the time for the first practical application of the electron began, the 19th century delivered to the world the **incandescent light bulb,** and the invisible became visible.

Two centuries have passed, and **electrical power** changed the world as no other force did.

We cannot see our energy either. But it is there.

A brief introduction about who is giving you all this information

My parents met on the 6th of December 1989. They were of different culture and different religion.

After one year and one day, I was born.

Years have passed and I was growing.

Curiosity started to develop. My questions evolved to a more complex degree. I quickly felt in love with the language of the universe. With math. It offered answers.

My father believed in the teachings of sports. In my young years, he chose what sport I should do. I trained 2 years in gymnastics. There, I understood what true physical strength means. But the hasty training, at a young age, in that field, could negatively impact my physical growth process. My father decided the field of athletics would be better. 3 years of training and I understood what resistance meant. I had potential. Something needed to be done with it. I started professional football. I stayed there 5 years. I had the training; I had the knowledge, but there was something lacking. This is a team sport. I felt that there was no team. The conflict was within the team. I had no desire to fight for a cause that did not have strong values.

So, I started deciding what sport I should do.

In my adolescent years, I found a trainer and I asked him to teach me martial arts. I rapidly adapted to the sport. I had strength; I had resistance; I had the will. I also had the desire to understand how I can become greater! So, I did the math. The answers were not coming from the exterior. I started to become aware of a hidden force.

I discovered myself and I started analyzing everything!

I learned how to control and tame my energy.

That was a mistake.

I had become a calm, patient, rational person. My energy was stable. That meant progress would have stopped. And chaos eventually appears.

My father had battles.

Corruption was the main antagonist in my country in the early years of 2000. My father was a victim of it, but he did not bend the head. He decided to fight the system. You cannot win a war if the rules are made by the adversary. He did not understand this. He lost the fight. The financial cost of the battle has left our family without a home and my father was put in jail.

I would have become the one making the decisions in my father's family. My primary desire was to offer wellbeing and financial support.

I had resolved most of the problems. My math abilities served me well into analyzing a problematic event. Into detecting its source and resolving it.

But problems would still appear. I needed to understand the root cause, so I could end them!

So, I meditated. I searched for the exterior problem, inside. I would question the cause of events. I would enter the detail. Doing so, the clarity of the information would become dimmer. The brightness and clearness of the information would become darken. Shadows would appear. I would go deeper. Until there was nothing. Complete darkness. I entered chaos! It felt as if my entire mind was collapsing. The structures holding the information had no meaning. Every thought started deconstructing itself. Every notion screamed of confusion. Nothing made sense. I had no way out. There was nothing to pull me back. I did not know where I was going. I did not put any "anchors".

Until something awakened. Something tamed by me many years ago.

In all that chaos, in all that noise, something emerged from within and silenced all the screaming noises! A powerful red light arose, and everything went quiet. Nothing dared to challenge it.

I had reached peace. I had reached serenity. Everything was calm and I was stable. Unfluctuating and undisturbed. An immense flow of energy was passing throughout my entire body. It felt powerful. It felt unnatural.

That was the first time I had become fully aware that there is something inside that protects me.

I was captivated by that power, and I knew that I must keep it!

So, I started to train myself into understanding this energy. I know what triggers it! It was triggered by darkness. And in the darkness, I understood light!

I learned how to increase the amplitude and frequency of my energies. To use them! To begin to grow again!

And use it to solve problems!

My name is **Cătălin-Nicolae FARCAȘ**. I am an electronics and telecommunication engineer and I think I found out how the spirit works.

What is the spirit?

The complete harmonious communion between electrons forms the energy structure inside us known as the spirit. It behaves like an electric current and therefore adheres to the established conclusions regarding electrons.

It is our life source. In it lies our behavior, energy, and identity.

To understand our spirit, we need to understand what lies in it. Let's start with the behavior.

Our behavior is not defined by our vessel. **We are not our body**. Our identity is not determined by our physical characteristics, such as height, weight, hair color, eye color, skin color, or genitalia. **We are not our mind**. Our identity is not determined by our mental processes, such as throughs, way of thinking or cognition. **We are our spirit**.

To gain a greater understanding of the hierarchy of **body**, **mind**, and **spirit**, we need to comprehend their respective importance.

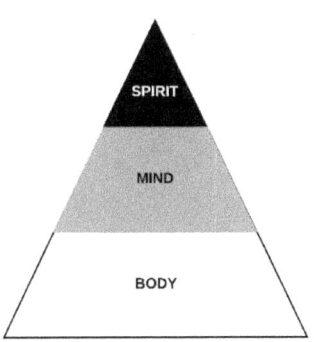

The spirit, although invisible, holds the utmost significance. Following that, the mind takes precedence as the second smallest and most important aspect. Lastly, the body holds the least importance. The mind surpasses the body in importance due to its swiftness. Mental efforts are quicker than physical efforts. Similarly, the spirit surpasses the mind in importance because it operates at a higher speed. **Actions** and **reactions** occur more rapidly than thoughts.

This would be the answer to the behavior. Our behavior is shaped by our **actions** and **reactions**. Each action and reaction we engage in is fueled by **energy**. This energy flows throughout our body, and it is directed to the intended action or reaction, resulting in movement, or thought processes.

Underneath every action and reaction, there exists a motive. A reasoning why they have been executed. Actions and reactions have a source of beginning. And in that source lies our **identity**.

Our spirit is a structural energy that is formed by three things.

- The SPARK
- The WILL
- The THREE ENERGIES

The **spark** is our identity. It describes our <u>truest, purest, and uninfluenced self</u>. Our spirit is like a "fiery water" and behaves like an electric current. It is an energy structure that travels throughout our body. When used to an intense degree, it can heat our body to a certain point. It is like a fire inside and all fires ignite from a spark.

The **will** is our action. It is our <u>ability to make decisions</u>. It is the energy that resonates to an entity and acts towards it. It is our only force that can create action.

The **three energies** are our reactions. They are individual different energies that when they react, they grant us energy. They will be presented as the RED ENERGY, YELLOW ENERGY, and BLUE ENERGY.

THE SPARK

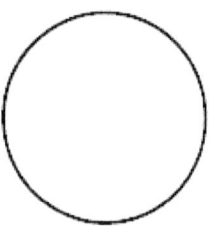

Understanding the spark

Every living being enters this world, through a form of creation, naked. Without anything to its possession. With no evil thoughts. With no altruistic thoughts. Without a developed mind. It enters and starts its journey into this world **from zero**.

Even us Humans, from birth we do not understand the world we live in until we developed a mature mind and start to use it to analyze the world and ourselves.

Until we become aware that we have a spirit inside, we journey through this world as a **mirror spirit**. A mirror spirit is a living being that studies other living beings similar to its own species and replicates or reflects the behavior it resonates with.

If we want to be as our father or mother, we study them, admire them and we borrow their behavior as ours! We emulate their actions and characteristics as a way to shape our own behavior and identity. <u>We mirror them</u>.

As we grow older, we might find other possible role models and mirror that behavior as well. We become a product of influence and if we do not search for our own spirit, we will not discover our truest, purest, and uninfluenced **self**.

The most common and influential factors that shape a person's development and behavior are <u>family</u>, <u>education</u>, <u>social environment</u>, and <u>life circumstances</u> Our family, as the first social unit we encounter, plays a significant role in shaping our beliefs, values, and attitudes. Education provides us with knowledge, skills, and perspectives that influence our thinking and decision-making. Our social environment, including friends, peers, and colleagues, can greatly impact our beliefs, behaviors, and aspirations. Additionally, the circumstances we encounter throughout life, such as challenges, opportunities, and experiences, shape our perspectives and choices.

These influences are layers that enclothe our spirit. Our identity lies deep hidden in our spirit. In order to discover our spirit, we must first view our current decisions. We must analyze if our decisions are a product of influence. We need to reevaluate them in order to ensure that they come from our own desire. If not, we will remain a mirror of someone else. Our decisions must be true. They must be pure and influenced.

When our decisions are pure and influenced, we can study them and discover our identity. Our identity is reflected in the decisions we make. Our decisions are guided by our **set of values**. To better understand what are values we first must understand what is important.

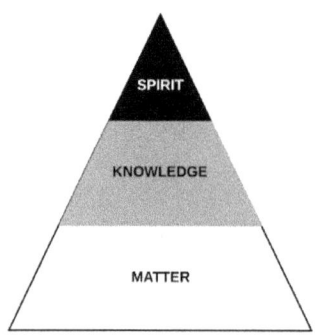

The most important aspect of our existence is our spirit, as it gives us life and vitality. Without it, our bodies would be lifeless vessels. Following that, knowledge plays a significant role in our lives. It allows us to gain understanding of the world and ourselves, enabling personal and intellectual growth. Lastly, matter holds its own significance as it provides us with resources and materials that we can utilize for various purposes.

In order to secure and protect what we believe to be important we need something that does that. We need values.

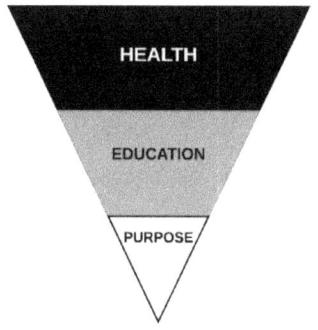

These are the main values. The first one is health. It is the most valuable one. When we have it, almost anything is possible. The next one is education. With it, we can decipher any mystery we encounter. The last one is the purpose. It is the main direction of the spirit. It is placed on the last place because in order to achieve it we need education and without health we will have nothing.

The identity of a living being is the **spark**. The spark **is the sum of all decisions**. Our decisions are guided by our set of values. The set of values secure and protect the entities we find important.

In the deepest part of our spirit, in our **spark**, lies our set of values. They are the things that secure and protect the entities we resonate with. **Our connections.** The deepest and most important connections are spiritual. Based on what we value the most defines our spark. If we value ourselves more than others, we are selfish. If we value others more than ourselves, we are selfless. A balanced spark values each life. It values its neighbor and itself!

If we do not value spiritual connections, we are neutral. If we value education the most, we are knowledge seekers. If we value material possession most, we are materialists.

The question about identity, of the **spark**, is "Who am I?". The answer is rather a simple one. We either are **selfish**, **selfless**, or **balanced**.

Balanced spark

The balanced spark is neither selfish nor selfless. It holds value not only for itself but also for others. It values every living being, recognizing the worth of each individual as well as its own.

The balanced spark recognizes the inherent value of every living being and also acknowledges its own worth. It understands that all life forms have a unique place and purpose in the world. By valuing each living being, the balanced spark embraces the interconnectedness and interdependence of all life. This balance between valuing others and valuing oneself is crucial for maintaining a harmonious and compassionate existence. The balanced spark strives to create a world where all living beings are respected, valued, and appreciated, while also acknowledging the importance of self-love and self-expression.

Achieving a balanced spark can be one of the most challenging endeavors in life. Being a selfish spark can prioritize your preferences and interests, but it also can strain relationships with others and lead to conflicts and isolation. Being a selfless spark can strengthen relationships and foster a sense of trust, support, and camaraderie, but it also can compromise your preferences and interests.

The universe has a way of humbling those who are excessively selfish and can lead to feelings of resentment for those who are overly selfless. Because we are intricately connected with the universe. Electrons are the hidden life force of the universe. They are omnipresent and can be found in every corner of our existence. Electrons are the hidden life force of ourselves. They travel inside the very atoms of our body and form our spirit. Through quantum phenomena, electrons have the ability to become entangled as a result of their interaction with electromagnetic forces. Thus, there exists a profound interconnectedness among all things, and they are composed of energy. The purpose of energy is to achieve stability, allowing it to endure for an extended duration. To maintain this stability, the universe responds to any actions and reactions that are imbalanced by readjusting itself.

Every action and reaction you create has an echo. It reverberates back to restore balance in the universe.

It is essential to achieve a balanced spark, as the universe will reflect and echo your actions and reactions back to you until circumstances balance you.

The spark serves as the essence of the spirit. The teaching of the **balanced spark** can be subtly observed as the foundations of all major religions. The teaching of the balanced spark is known as the **golden rule**.

Christianity
"Do unto others as you would have them do unto you."

Islam
"None of you truly believes until he wishes for his brother what he wishes for himself."

Hinduism
"This is the sum of duty: do not do to others what would cause pain if done to you."

Buddhism
"Hurt not others in ways that you yourself would find hurtful."

Judaism
"What is hateful to you, do not do to your fellow man."

Sikhism
"Treat others as you would be treated yourself."

Confucianism
"Do not do to others what you do not want done to yourself."

Jainism
"A person should treat all beings as he would himself like to be treated."

Bahá'í Faith
"Lay not on any soul a load that you would not wish to be laid upon you."

Zoroastrianism
"Do not do unto others whatever is injurious to yourself."

Taoism
"Regard your neighbor's gain as your own gain, and your neighbor's loss as your own loss."

Native American spirituality
"Respect for all life is the foundation."

African traditional religions
"What you wish your neighbor to do to you, do it to him."

The question about identity, of the **spark**, is "Who am I?". The balanced spark removes the question because it does not compare itself with anyone. Because everyone is of value.

When the question about identity fades, another, more important question appears.

The question of the **will**.

THE WILL

Action

The spirit is our life source. In it lies our <u>behavior, energy</u>, and <u>identity</u>. In response to the question of identity, "Who am I?" the **spark** serves as the ultimate answer. Our behavior is shaped by our **actions** and **reactions**. Each time we act or react, electrons traverse from one atom to another, providing us with electrical energy. Therefore, with each action and reaction, we are immersed in abundant **energy**.

The **will** is our action. It is our <u>ability to make decisions</u>. It is the singular energy force within our spirit that can create movement, thinking, and action.

It starts with **intention**, and it resonates with the entity that you like. Then you use this energy to initiate the action to connect with that entity.

The **will** is driven by the answer to the question "What do I want?". Wanting is the first step, in acting. It is not enough to want something. If you want it, you must take decisive action to pursue and attain it.

When we do not know what we want we act and remain as a <u>mirror spirit</u>. We will journey through our whole life and make decisions based on "I have to" statements.

When you pose the "What do I want?" question, an electric response is received. When applying signal processing to the electrical response, it is possible to decode a message, which the mind then interprets as a voice. When you ask yourself more and more "What do I want?" questions you will start to perceive and access your **inner voice**.

The most important thing is the spirit. The singular energy force of the spirit is the **will**. The will is driven by your inner voice. Therefore, the most important thing to the spirit is your word! When you speak it, your interior is listening, and if your word is worthy of respect, your reactions will respect it and they will fight for what you want.

Inner voice

The spirit is a structural electrical energy. The flow of electrons within the intricate structure of our body traverses every atom, giving rise to remarkable abilities such as movement, thinking, and action through electron-based **communication**. Using signal processing we can decode a message. That message can be interpreted by the mind as a voice. We can access our own spirit's electrical response and discover answers about ourselves. We can use our mind to communicate with our own spirit.

Questions serve as the paths that lead us to answers. When we pose questions to ourselves, we receive electrical responses, and within those responses, the answers reside.

By posing specific questions, we elicit precise signal responses. When we ask questions about our desires, we access responses from our **inner voice**, which represents the voice of our will.

The inner voice holds paramount importance as a guiding force.

By enabling access to our inner voice, we can engage in three practices that strengthen our will.

- Meditation
- Balanced spiritual state
- Discipline

Meditation

Our spirit is a structural energy that is formed by three things. Our **spark** is our identity. From it our **actions** and **reactions** emanate, defining who we are and how we interact with the world. Our <u>inner voice</u> lies in our action. Our **reactions** have their own voices as well.

There is a force and three energies inside us. There are four voices inside us.

Once we have accessed our **inner voice**, we can use a powerful technique in which we can form **communion** with our **reactions**. This technique is called **MEDITATION**.

When we ask questions to ourselves, we receive electrical responses. Those electrical responses are decoded by the mind as words. When we ask questions to ourselves, we receive answers.

When we ask what our fears are, we receive answers. When we ask what our angers are, we receive answers. When we ask what our faiths are, we receive answers.

Communion happens with <u>communication</u>. And can only be done through <u>meditation</u>. Communication will strengthen your voice and communion will strengthen your will.

Meditation is powerful due to its ability to reduce stress, improve focus, enhance emotional well-being, increase self-awareness, promote mental and physical health, and cultivate mindfulness, benefiting overall well-being.

Balanced spiritual state

Our current individual inner and present condition is our spiritual state. The **spiritual state** lies in our action. In our <u>will</u>.

Our willpower is determined by our spiritual state. If we are experiencing feelings such as depression, sadness, boredom, etc. it is a sign of low willpower. A sigh of high willpower is peace, happiness, purpose, etc.

The essence of life exists within all living beings, whether they are plants, animals, or humans. Each displays actions and reactions that indicate the presence of a spirit. Let us pause and reflect on dogs for a moment. They seem to have discovered the secrets of life and eternal happiness. That is because they live actively in joy in the present moment. So, what might be the secret of happiness?

"The secret of happiness" is **gratitude**.

Gratitude allows us to find contentment and joy in the present moment. However, as humans, our experience of happiness is different from that of dogs. Our minds have the remarkable ability to divide our current state into three distinct channels. With our advanced and highly developed minds, we can recall specific moments from our past and envision scenarios that lie far ahead, surpassing the predictive capabilities of animals. However, it is in the present moment where gratitude resides. Unfortunately, our minds often trap us, pulling us away from the present. Our memories can anchor us in the past, while our imagination can propel us into the future. As a result, our present lives become divided into the realms of **past**, **present**, and **future**.

Due to this division, our spiritual state is divided as well. In order to achieve a balanced spiritual state, the past needs **peace**, the future needs a **purpose** and the present needs **gratitude**.

Let's start with the past. We all begin our journey into this life from zero. In our process of growth, we might have done some things that we are not proud of or are ashamed of and now our memory disturbs us. What is the solution? The past cannot be changed. How do you cope with something that you cannot remove, and you have to live with it?

The solution is **peace**.

Let's imagine you have stolen something. Something from another person. It does not matter the reason for it, you have stolen it! It is a selfish act, and your reactions start to speak to you. You will hear voices such as "thief", "loss", "unworthy". These words undervalue your **spark**. The past cannot be changed but you need peace. Peace is a negotiation with a resolution. You need to negotiate with your other voices. You need to negotiate with yourself and make a statement. A statement that will resolve your future actions. The future can be changed! You will say, "I do not want to be a low value spirit". "I want to be high value one". "**I will never steal again**!" – This is your resolution! Your reactions will not react to this statement because it is desired by the balanced **spark**! If you do not steal again, your reactions will not react, and you will have peace!

Peace is a responsibility not a realization. You must be true to your statement or peace will decrease in value and the past will disturb you again. Honor your word and you will have peace.

Let's move to the future. The spirit is a fiery water, and it behaves as an electric current. An electric current needs a direction in order to flow. It needs a **purpose**! Using the ability of the mind, using imagination, you can create a vision! And you can lead yourself towards it! Without a direction, the spirit will be condemned to live a life without purpose. A meaningless life.

When the past is at peace and the future has meaning, what remains is just **to enjoy life**!

Gratitude is experienced in the present moment, and cultivating it requires recognizing that nothing can be taken for granted. Nothing truly belongs to us, as we enter this world with nothing and leave with nothing. Our only true possession is our physical vessel. Everything else we encounter in life are temporary gifts. Clothes are temporary. Toys are temporary. Friends are temporary. Parents are temporary. By gaining an understanding of the dark truth, we develop a deeper appreciation and begin to value things and people more. Darkness empowers our **will**.

Having either peace, gratitude, or a purpose can bring us happiness. Having all three brings us into a **balanced spiritual state**.

With a balanced spiritual state, our **will** becomes free! We liberate ourselves from the influence of external forces and we can act as we want.

Discipline

The most potent and efficient method to enhance one's willpower is through discipline. Once you have accessed your **inner voice**, you can establish a set of rules within yourself to guide your approach to the known world.

- In the morning, first thing **I will** drink a cup of water.
- Then **I will** make my bed.
- **I will** brush my teeth.
- ...

What are these? These appear to be statements. Where do they come from? From yourself. They are statements from your inner voice. If the statement can be executed, the **will** is strong. Honoring your statement is the most stable thing you can do for your spirit. It helps yourself and it helps others.

Discipline is setting a rule by yourself and respecting it. A rule creates a habit. Many healthy habits can create a healthy life. A healthy body, a healthy mind, and a healthy spirit. When your spirit is in a low willpower state, your habits will rise you! Because the will was, for a long time, embedded into your vessel. It will be easy to fulfil your self-rules even in your low willpower state. Your habits will save your spirit.

THE THREE ENERGIES

Everything starts with belief

Belief is the bridge between the mind and the spirit.

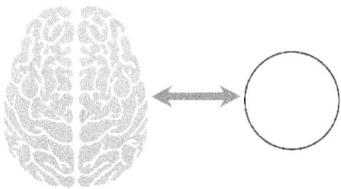

Belief is something that is neither part of the spirit neither part of the mind. It is part of both. It is as a bridge that connects them together. Without this connection, our vessel would be without a spirit and our spirit would be without a vessel. Belief is not a component of the spirit, but the spirit reacts based on it. Belief is not a component of the mind, but the mind needs it to form points of beliefs.

Belief is a **measuring instrument**. It subjectively measures the acceptance degree of the perceived **truth**!

Our sensory input originates from our senses: the eyes transmit signals to the brain, interpreted as images; the ears convey signals, interpreted as sound; the nose detects and sends signals, interpreted as smell; the tongue relays signals, perceived as taste; and the skin transmits signals, interpreted as touch.

Based on the sensory input, **belief** subjectively measures the acceptance degree of the perceived truth. It measures reality – the **exterior world**; and transforms and creates our **interior world**.

The measurement is done in percentage.

The mind is a map of points of beliefs. When our **belief system** subjectively measures the acceptance degree of a perceived truth, it transforms it into a <u>point of belief</u>.

Points of belief are between 1 and 99%. When belief is 100% or 0% it is not belief anymore. It becomes **certainty**. Certainty is a blocking point in the mind.

The spirit is a structural electrical energy. It travels throughout our whole body and our whole mind. When the spirit encounters a point of certainty, it cannot move through it and the spirit remains blocked. When darkness comes, our spirit will not be able to reshape that point quickly and adapt to change because that point has been consolidated.

Points of belief are between 1 and 99%. They are not fixed points. That means those points can fluctuate. That means our spirit can fluctuate – it can move! If our mind is composed of blocking points, points of certainty, the spirit cannot move. We will remain the same. In times of peace, we are calm. In times of darkness, we are doomed.

Points of belief make us an ADAPTOR!

The only things that you can trust to be certain are **darkness** and the things that you have **witnessed**.

Reactions

The spirit has only **one action** and **limitless reactions**. Reactions can be of different intensities and are all felt different. When reactions occur, they bring a gift. They grant energy. They bring a tremendous contribution that enforces the spirit's power.

While the number of possible reactions is limitless, they originate from **three primary variants**, from which they ramify and form a multitude of possibilities.

We have three primary reactions:

- The RED energy
- The YELLOW energy
- The BLUE energy

A reaction is an action executed in response to a situation or event. It is considered an action because the **will** executes the action due to it being overpowered by the energy of that specific reaction.

Reactions are a product of one's own **belief**. They cannot exist without belief.

To gain a deeper understanding of reactions, it is essential to comprehend their underlying **causes**, which originate from various situations and events.

Situations & events

Our world of matter, energy, and forces is in constant motion. Everything is interconnected, and movements in one place can eventually give rise to situations or events elsewhere.

Situations and events come in a multitude of forms. They may range from simplicity to complexity, from the pleasant to the unpleasant.

Our world is characterized by a constant stream of situations and events, and this dynamic will endure indefinitely. They can range from simple and pleasant as a dog smiling at us or complex and unpleasant such as a global crisis. They will continue to occur due to the first law of the world – **that everything changes**.

Despite their diversity, situations and events can be categorized. Situations and events can be categorized into three distinct groups.

The 3 black dots of darkness

Darkness is the layer behind our world of matter, energy, and forces. It encompasses and envelopes the world inside it. We do not know what it is, but we found some traits of it. The physical realm is incapable of detecting it; we cannot see, hear, touch, smell, or taste it.

Only energy possesses the sensitivity to detect it, and it responds instantaneously to this detection. Our spirit reacts to it as if it were designed to be the perfect counterforce to darkness.

As if our world would be a fiery water and darkness would be a viscous oil. They do not mix. They are as immiscible liquids. They are two separate things of the same universe.

Universe	
World	Unknown
Energy	Gravity
Expansion	Contraction
Light	Dark

When one moves the other responds.

Darkness is the empty galactic void that exists "outside". We do not know what it is. But it is empty. It is quiet. It is cold. And it gives nothing. Due to its contraction traits, darkness sometimes succeeds to infiltrate into our world.

It is subtle and it can breach our world through three paths. It starts unnoticeably and moves the molecules of the world creating a chain of situations or events.

Darkness hides in events and situations.

Our physical senses can detect it through our physical world. They can detect some seismic activities. Some deformations that might happen. Anomalies that can be witnessed. Until **CHAOS** emerges! It is unexpected, and it brings panic! The world surrounding it trembles around it. If it is left unchallenged, chaos can bring **LOSS**! And finally, it can bring **DEATH**!

There are only three certainties in the world. Only three 100% beliefs. Three unmovable things in our moving world! They are known as the three sacred truths. The three DARK TRUTHS!

They are the first laws of the universe.

They are called the 3 black dots of darkness.

Darkness can be categorized into three distinct groups.

- Chaos
- Loss
- Death

The 3 black dots of darkness are the **causes** that creates situations and events.

Darkness can infiltrate in our world, but it cannot destroy it. Due to its characteristics, the first laws of the world emerge:

1. Darkness does not destroy. It changes.
2. Change creates problems and solutions.
3. Every problem has a solution. Every solution has a problem.
4. The solution lies in the problem. The problem lies in the solution.
5. There is always a solution. There is always a problem.

Darkness has its laws, and the world shapes its laws based on darkness. Because we are inside it. Deep down if we dissect every possible existing event or situation, we will encounter darkness.

There are two types of darkness:

- Manifestations of darkness
- Pure darkness

The manifestation of darkness can be detected by our physical senses. Darkness can manifest into smaller scales such as illness, petty conflicts, violence, or larger scales such earthquakes, volcano eruption, tsunamis and much many more. The 3 black dots are embedded within them.

Manifestations of darkness can be witnessed in the exterior world.

Pure darkness is the 3 black dots themselves. The presence of darkness resides within you. When the exterior world is at peace, but you are agitated inside, there is pure chaos inside you. When a loved one has passed away, but you continue to feel loss, there is pure loss inside you. When you are intimidated by the incoming end of life, there is pure death inside you. The 3 black dots are embedded within you.

Pure darkness can be witnessed in the interior world.

The manifestation of darkness

Our world of matter, energy, and other forces is founded upon a layer of dark unknown.

Due to its contraction traits, darkness can influence our world.

Let's say a pandemic happens.

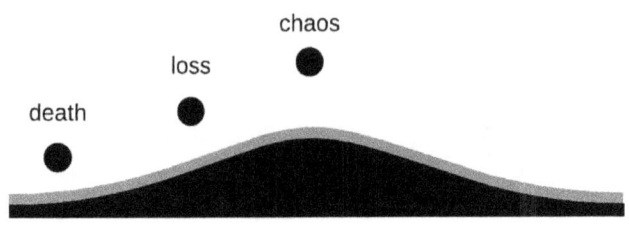

As a result of the influence of darkness, a series of biological events is triggered, leading to the emergence of a deadly disease.

When darkness contracts and influences our world, it results in a noticeable curvature, often represented as a **bulge**. This curvature's impact on our world gives rise to distinct **elevations** and **depressions**.

The bulge creates RISE and FALL edges.

At the zenith of the curvature, Chaos manifests.
During this time that people become aware of the pandemic.

Is chaos a bad thing? Is it a good thing?

Bad for	Good for
Human Health	Ecological Environment
Restaurants	Pharmacies
Offline Businesses	Online Businesses

When one thing rises, another falls, resulting in the curvature.

Chaos does not destroy. Darkness does not destroy. Due to this statement: **Every problem has a solution.**

Let's continue.

At the midpoint of the curvature, Loss manifests.
This is when some individuals begin to lose their jobs due to the impact of chaos and the inability to control its outcomes.

As people begin to pass away, it becomes evident that all 3 black dots are interconnected, each one seemingly attempting to bring out the other.

At the lowest edge of the curvature, Death manifests.
You become infected. Gradually, you feel death starting to embrace you. You have fever. You are breathing heavily, sweating profusely. Your usual problems lose their grip on you. Nothing else matters. Your sole focus is on survival.

It is a slow, relentless process. You feel the grip of chaos, loss, and death tightening. Your world has transformed. Friends are gone, and you are slowly losing your life. The reality of death becomes palpable. Darkness attempts to reduce everything to emptiness, silence, and coldness, until you feel nothing at all.

Fortunately, your immune system has saved your life. You have lived to see another day, but you now know how darkness feels.

During the pandemic, some people felt chaos. Some felt chaos and loss. Some felt only loss. Some left chaos, loss, and death.

For some, chaos appears as something unnoticeable. It is almost imperceptible, occurring daily and becoming a part of our normal lives.

The 3 black dots of darkness are hidden in all situations or events. It does not matter if its scale is large or small or its intensity is high or low, darkness affects energy thus induces reactions.

Chaos

The manifestations of chaos encompass numerous representations, including disorder, confusion, unpredictability, the absence of a solution, uncertainty, the unexpected, turbulence, luck, anarchy, noise, mayhem, complexity, unrest and the most known of them all, **randomness**.

When the manifestation of chaos occurs, the scope of darkness is to introduce pure chaos into one's spirit. Darkness can introduce pure chaos with its common tactic known as **surprise**.

Once pure chaos has infiltrated one's spirit, and depending on the intensity of it, darkness can incite **PANIC**. Panic serves the role of agitating the reactions inside one's spirit.

Loss

The manifestations of loss encompass numerous representations and are structured by a hierarchy of importance. They include <u>material loss</u>, such as the losing of a pen, a wallet, a house, or a job, as well as <u>loss of knowledge</u>, such as memory loss, amnesia, forgetfulness. Additionally, there is <u>spiritual loss</u>, such as the grief of losing a friend, a sibling, or a parent, and most commonly, the **loss of time**.

When the manifestation of loss occurs, the scope of darkness is to introduce pure loss into one's spirit. Darkness can introduce pure loss by disrupting an existing connection between one's spark with an entity.

Once pure loss has infiltrated one's spirit and depending on the intensity of it, darkness can incite **AGONY**. Agony serves the role of agitating the <u>will</u> inside one's spirit.

Death

The manifestations of death encompass numerous representations, including disease, threat, conflict, bullying, aggression, discrimination, harassment, crime, stalking, oppression, maltreatment, persecution, challenge, tyranny, punishment, torment, torture, all subjecting one's **mortality**.

When the manifestation of death occurs, the scope of darkness is to introduce pure death into one's spirit. Darkness can introduce pure death with its common tactic known as **intimidation**.

Once pure death has infiltrated one's spirit, and depending on the intensity of it, darkness can incite **DESPAIR**. Despair serves the role of agitating the spark inside one's spirit.

Two worlds

The spirit operates in two environments. There are two worlds:

- The exterior world
- The interior world

The **exterior world** is where darkness and real world exist. It is called reality. Your power of control is limited by your environment and your vessel's physical capabilities. It is the same world where your friends, loved ones and all the living beings live. It is <u>only one</u>. The laws of universe govern that world. You can influence that world with your words or actions.

The **interior world** is where the spirit lives. Every living being has its own unique interior world. You have an interior world, and it is completely yours. Exterior factors might influence it, but the control is yours. You govern your world. There are no laws. You can shape it as you desire. You can control your interior world with the interior tools.

Interior tools

The interior world is the home of the spirit. In one's early stage of development, the spirit measures, with its belief system, the exterior world transforming and creating its interior world. Based on the maturity of the spirit's perspective of time and space its interior world would be calibrated accordingly.

When one becomes aware of its interior world, it can use three main interior tools to shape it in alignment with its desires.

- **Intention**
- **Will**
- **Imagination**

Creation requires a starting point, an end point and what remains is for the spirit to form the structural connection between those points.

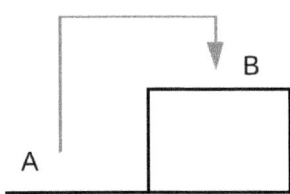

In order to get from A to B, the **interior tools** are required.

The spirit uses **intention** as a starting point. It uses **imagination** to create a destination as an end point. And it uses the **will** to form the structural connection between those points.

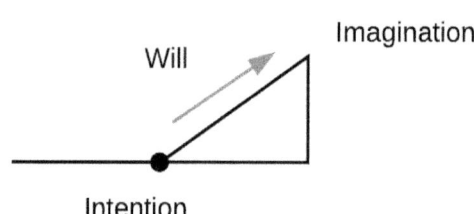

Through intention, will, and imagination, one can create anything. Initially, in the interior world, and subsequently manifest it in the exterior world, thereby influencing and reshaping the very reality that encompasses all life.

Colors

One way to gain insight into the workings of the world is to comprehend its intricacies through the study of colors.

There is an interconnected relationship between <u>colors</u>, <u>light</u>, and <u>energy</u>. Colors are a manifestation of light and light is a manifestation of energy.

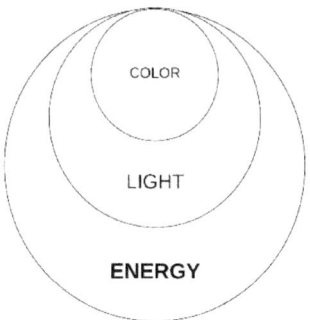

Darkness and **light** are the extremes of the universe, akin to **black** and **white** representing the endpoints of the color spectrum.

Black is the absence of all colors, making it a <u>non-color</u>. When all colors are absorbed and no light is reflected, the result is black.

White is the combination of all colors, earning it the designation of an <u>all-color</u>. When all colors are combined or reflected, and no color is absorbed, the result is white.

Darkness is the absence of light. When light or white undergoes de-structuring, it reveals the full spectrum of colors. The spectrum of colors originates from the synergy of three primary colors.

THE RED REACTION

The red reaction is one of the three primary reactions of the spirit.

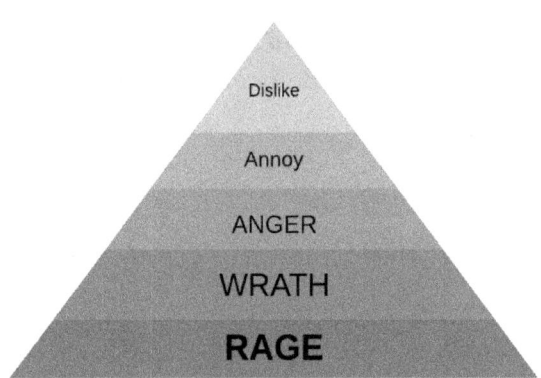

The RED reaction is the <u>belief</u> that something **changes**.

Reactions respond to the presence of darkness and its manifestations. Each time we react, the spirit traverses from atom to atom in our vessel, generating energy. Therefore, with each reaction we are immersed in abundant energy.

It is the quickest and most intense energy in the spirit. As soon as darkness appears, it reacts. When change occurs, the intensity of the red energy is directly linked to the level of darkness. In other words, if the intensity of darkness is high, the intensity of the red energy is also high.

The **red reaction** responds the outcome <u>change</u> of darkness.

The dynamic traversal of the red energy throughout our body is unmistakably felt and experienced as a distinct sensation of **dislike**.

As the intensity of darkness amplifies, so does the surge of one's red energy. This heightened energy manifests initially as <u>annoyance</u>, escalating to <u>anger</u>, further intensifying into <u>wrath</u>, and reaching its peak in a state of <u>rage</u>.

The red energy awakens, when darkness occurs, in order to protect you. This reaction serves the spirit with the role and tittle of **PROTECTOR**!

THE YELLOW REACTION

The yellow reaction is one of the three primary reactions of the spirit.

> **FEAR**

The YELLOW reaction is the <u>belief</u> that something **bad** is going to happen.

Reactions respond to the presence of darkness and its manifestations. Each time we react, the spirit traverses from atom to atom in our vessel, generating energy. Therefore, with each reaction we are immersed in abundant energy.

The **yellow reaction** responds to the possible outcome <u>danger</u> of darkness.

The dynamic traversal of the yellow energy throughout our body is unmistakably felt and experienced as a distinct sensation of **fear**.

The yellow energy awakens, when darkness occurs, in order to warn you. This reaction serves the spirit with the role and tittle of **WARNER**!

THE BLUE REACTION

The blue reaction is one of the three primary reactions of the spirit.

The BLUE reaction is the <u>belief</u> that something **good** is going to happen.

Reactions respond to the presence of darkness and its manifestations. Each time we react, the spirit traverses from atom to atom in our vessel, generating energy. Therefore, with each reaction we are immersed in abundant energy.

The **blue reaction** responds to the possible outcome <u>salvation</u> of darkness.

The dynamic traversal of the blue energy throughout our body is unmistakably felt and experienced as a distinct sensation of **faith**.

The blue energy awakens, when darkness occurs, in order to save you. This reaction serves the spirit with the role and tittle of **SAVIOR!**

Overview

<u>SPARK</u>

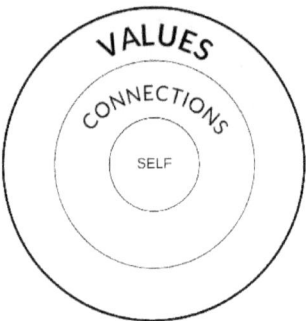

The identity of a living being is the spark. Our actions and reactions are guided by our **set of values** that secure and protect the entities we resonate with – our **connections**. The deepest and most important connection is with our own **self**.

ACTIONS & REACTIONS

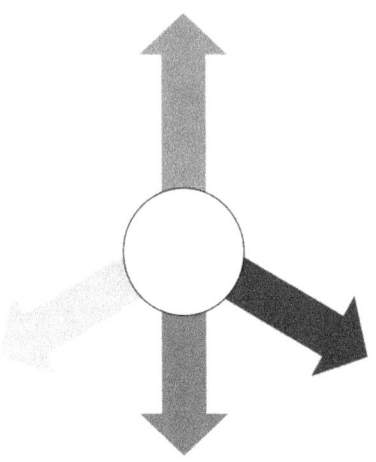

The spark is the sum of all decisions, from which actions and reactions branch out. The spirit has **one action** and **three primary reactions**. The <u>action</u> is the force of the spirit known as the <u>will</u>. It is the ability to make decisions. The three primary <u>reactions</u> are the energies of the spirit known as <u>fear</u>, <u>anger</u>, and <u>faith</u>. They respond to various situations and events.

SITUATIONS & EVENTS

Darkness is the unseen layer behind our world of matter, energy, and forces, encompassing and enveloping the world within. There are two types of darkness. **Pure darkness** cannot be detected by our physical senses – only our spirit can sense its presence. The **manifestations of darkness** can be detected by our physical senses, and they are represented through various situations and events.

EXTERIOR WORLD & INTERIOR WORLD

Acknowledging the existence and distinctions between the exterior world and the interior world, one can more acutely distinguish **reality** from **illusion**. The perception of truth will be more attuned with reality. The perspective within one's interior world has the potential to broaden.

THE SPIRIT

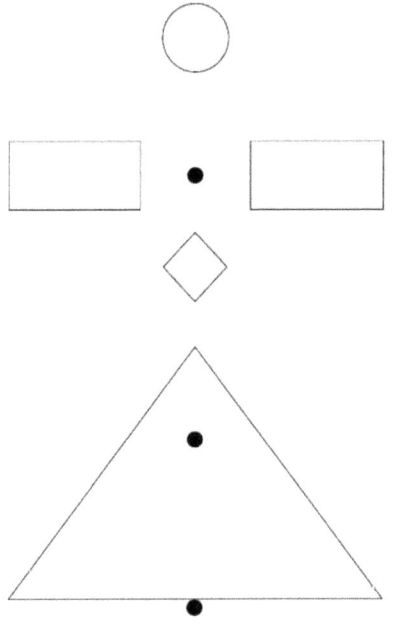

DARKNESS & LIGHT

Darkness is the vast galactic void that lies beyond our understanding. It is empty. It is quiet. It is cold. And it gives nothing.

We are currently experiencing a phase of rapid technological advancements in the field of energy – of light.

Light and darkness appears to be rivals.

In the absence of darkness, our spirit retains its primary force, derived from the spark.

This force holds the power to overcome any obstacle. It has earned the most esteemed title ever to exist. The will is the **chaos breaker.**

Structural representation of the spirit

The spirit along with darkness

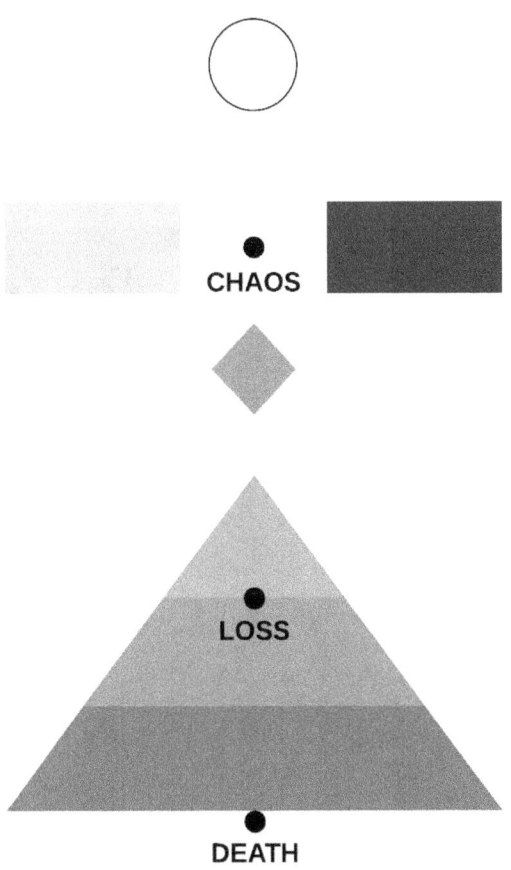

Spirit, darkness, belief, and a few emotions

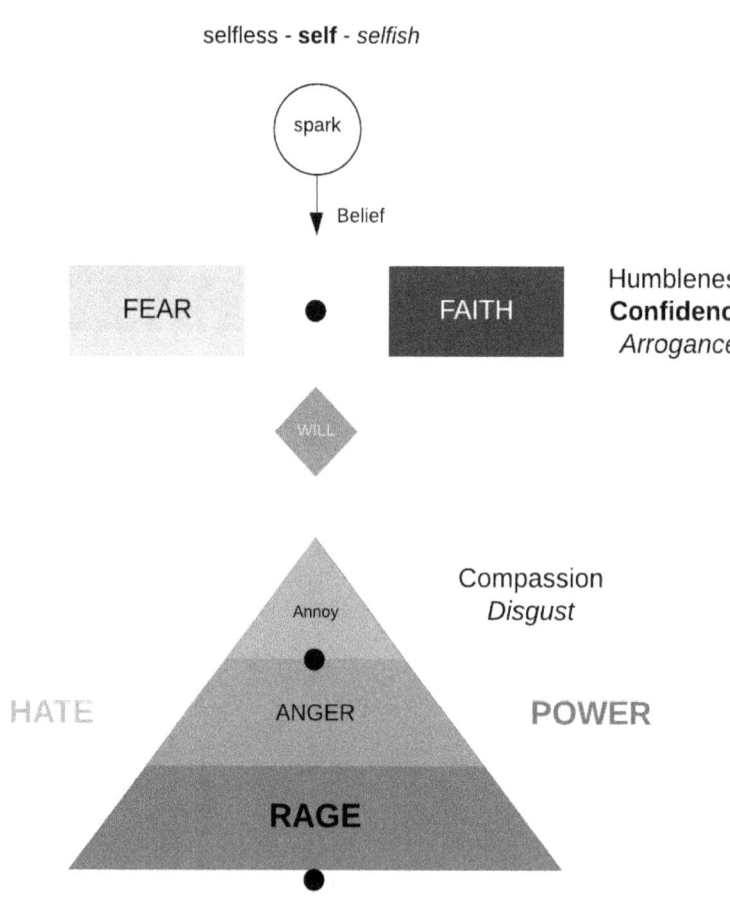

Superposition, fission, fusion

Energy can exist in <u>superposition</u>, undergo <u>fission</u>, and engage in <u>fusion</u>. It is imperative to discern its state for informed decision-making.

<u>Superposition</u>

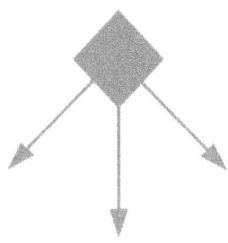

When our force exhibits multiple conflicting desires, leading to a state of contradiction, the will remains in superposition and unable to act. Superposition eventually leads to chaos. A decisive choice must be made.

Superposition can be experienced in both the force and energies of the spirit. When multiple conflicting reactions of the same energy occur, it is imperative to select only one so that the spirit continues its flow and does not remain stagnant.

Fission

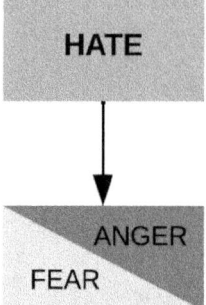

Hate is an energy formed by the combination of fear and anger. To address and resolve hate, it is necessary to fission it into its constituent elements – fear and anger. By challenging these energies separately, we can effectively put hate to rest. An effective approach involves transforming fear, which identifies the problem, into faith, which discerns the solution.

Fusion

<u>The Blue energy:</u>

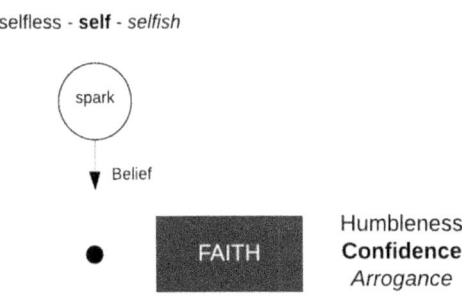

The manifestation of the blue energy through the spark results in various emotions, contingent upon the nature of the spark.

In the presence of the blue energy, a selfless person experiences the emotion of humbleness, while a selfish person feels the emotion of arrogance. Despite their apparent differences, both humbleness and arrogance originate from the same energy.

Humbleness and arrogance share similarities with confidence, as they all stem from the same emotional source – the blue energy. Despite their apparent differences, these emotions evoke similar feelings.

The Indigo energy:

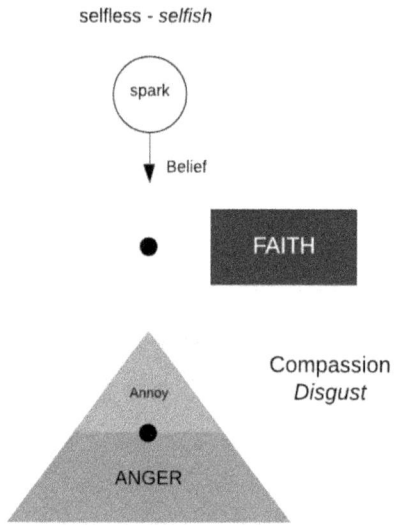

The fusion of blue energy with a slight of red energy forms the **indigo energy**.

Depending on the spark, there could be different variations of emotions. A selfless person, accompanied by the indigo energy, experiences the emotion of compassion. A selfish person, in the presence of the indigo energy, feels the emotion of disgust. Despite their apparent differences, both compassion and disgust originate from the same energy.

The Violet energy:

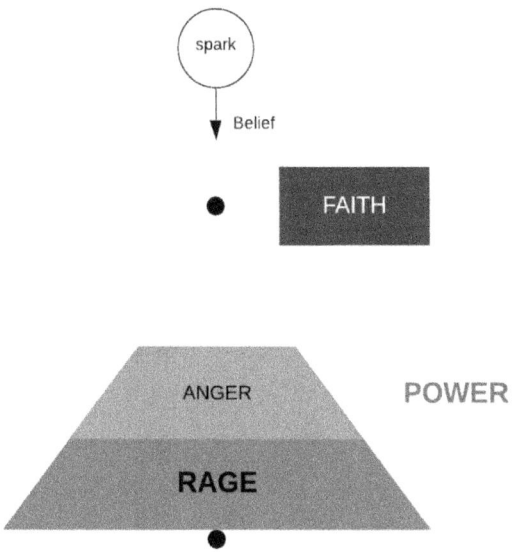

The fusion of blue energy with red energy gives rise to the **violet energy**.

The violet energy is the most powerful energy in the spirit, carrying a significant amount of both red and blue energy. The blue energy provides the direction, while the red energy efficiently filters out noise that may influence the direction. The violet energy represents a **powerful and focused signal**. Maintaining and controlling the violet energy is difficult and necessitates a strong and determined will for successful manifestation.

Mirror spirits

The quest for one's identity begins at birth. The journey of life starts from zero. Without a developed mind or body. Without knowledge of the exterior world and an empty interior world.

Until one matures enough to become aware of its life source, reflective behavior persists. Without self-guidance, individuals may shift focus from the interior to the exterior, mirroring <u>desires</u>, <u>dislikes</u>, <u>fears</u>, and <u>faiths</u>. To transcend this, one must venture into the interior world, discovering and awakening the voices of the inner self and warriors, breaking free from the state of a **mirror spirit.**

In the pursuit of identity and purpose, a spirit initially seeks answers in the exterior world. It mirrors desires, angers, fears, and faiths that are not inherently its own. True self-discovery commences only upon entering the interior world, marking the initial steps toward constructing one's <u>truest, purest, and uninfluenced self</u>.

Mirroring is a great ability that enables an intuitive understanding of a particular behavior without extensive mental contemplation. The refinement of this ability is recognized as **empathy**.

A positive aspect of being a mirror spirit is its attunement to common behavior, providing a sense of stability.

By relinquishing the stability of being a mirror spirit and exploring our inner selves, we evolve into free thinkers and free spirits, unlocking greater rewards such as purpose and destiny.

Connections

When we resonate with entities that evoke liking, disliking, fear, or faith, we form a **connection**.

When we resonate with something we **like**, we form a connection within our will, stored in our spark. When we resonate with something we **dislike**, we form a connection within our red energy, stored in our spark. When we resonate with something we find **bad**, we form a connection within our yellow energy, stored in our spark. When we resonate with something **good**, we form a connection within our blue energy, stored in our spark.

The most significant connections are those made through the will. When we powerfully resonate with something we like, it is called **love**. This powerful connection is intimate, and if privacy is not valued by the entities involved, the connection loses its intimacy.

A connection is a formed link between two or more entities.

Breathing

The spirit is the supreme controller of the mind and body; it can shape as it desires. Also, the body can have an impact on the spirit through breathing. The vessel can also influence the spirit into calming. When you do not have access to your interior tools, always remember that your vessel has one powerful ability to either calm or awaken your spirit. **It is the ability to breathe**. We do not breathe just to survive. We can increase, stabilize, or decrease the intensity of our spirit with a controlled breathing.

There are two main types of breathing.
- Chest breathing
- Diaphragmatic breathing (Belly breathing)

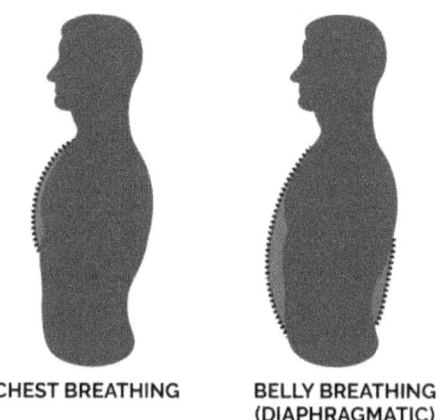

CHEST BREATHING   **BELLY BREATHING (DIAPHRAGMATIC)**

Diaphragmatic breathing is a more efficient technique which allows one to maximize the oxygen intake.

There are three actions in breathing.
- Inhale
- Hold
- Exhale

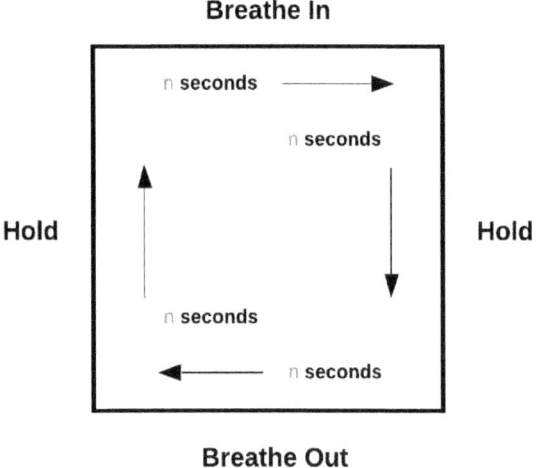

The variable "n" can increase or decrease the rhythm of breathing. For a harmonized frequency, this variable should remain constant in all stages.

THE THREE WARRIORS

There are 3 warriors inside us

Darkness has only one counterforce, and that is the spirit.

The spirit has **one action** and **three primary reactions**. The number of possible reactions is limitless; they originate from **three primary variants**, from which they ramify and form a multitude of possibilities.

Reactions are a response to a situation or event which are a manifestation of darkness. These reactions may manifest as <u>agitation</u>, <u>struggle</u>, or a <u>fight</u>. Our reactions have the limitless potential to fight everlasting and ever enduring battles. In their highest form of maturity, our reactions have the potential to bring the end of wars. Trained and matured, they are granted the tittle of WARRIORS.

Each one of us, each living being, has three warriors inside.

- The Red warrior
- The Yellow warrior
- The Blue warrior

The Red warrior

This warrior is known by the name:

THE
PROTECTOR

The **Protector** reacts to the outcome <u>change</u> of darkness.

The first line of defense against darkness is the red reaction. This is the most powerful warrior of the three.

In its immature and undeveloped phase, the Protector uses only one word – **NO**! In its mature and developed stage, the Protector has a higher level of fluency and precision in its dialect.

Each time we react, electrons traverse from one atom to another, providing us with electrical energy. Therefore, with each reaction, we are immersed in abundant **energy**. Applying signal processing to this electrical response enables the decoding of a message, subsequently interpreted by the mind as a voice.

The Protector <u>reacts</u>, <u>speaks</u>, and grants <u>energy</u>.

The Protector has a powerful ability.

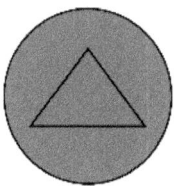

It has the powerful ability to cast **SILENCE**. It can say NO to darkness. No to panic. No to agony. No to despair. It can relinquish surprise and intimidation. It can say NO to the will and other reactions as well.

Without silence, there is only noise!

The Yellow warrior

This warrior is known by the name:

THE WARNER

The **Warner** reacts to the possible outcome danger of darkness.

In its immature and undeveloped phase, the Warner uses only one word – **BAD**! In its mature and developed stage, the Warner has a higher level of fluency and precision in its dialect.

Each time we react, electrons traverse from one atom to another, providing us with electrical energy. Therefore, with each reaction, we are immersed in abundant **energy**. Applying signal processing to this electrical response enables the decoding of a message, subsequently interpreted by the mind as a voice.

The Warner reacts, speaks, and grants energy.

The Warner has a powerful ability.

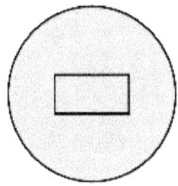

It has the powerful ability to cast **SAFEGUARD**. Without this ability the lifespan of every living being would be drastically reduced. Without this ability we would be dead.

Without safeguard, danger is imminent!

The Blue warrior

This warrior is known by the name:

THE
SAVIOR

The **Savior** reacts to the possible outcome salvation of darkness.

In its immature and undeveloped phase, the Savior uses only one word – **GOOD**! In its mature and developed stage, the Savior has a higher level of fluency and precision in its dialect.

Each time we react, electrons traverse from one atom to another, providing us with electrical energy. Therefore, with each reaction, we are immersed in abundant **energy**. Applying signal processing to this electrical response enables the decoding of a message, subsequently interpreted by the mind as a voice.

The Savior reacts, speaks, and grants energy.

The Savior has a powerful ability.

It has the powerful ability to cast **DIRECTION**. This ability bestows a path upon the will – a saving one! Your will must actively follow it for manifestation. Should you neglect this crucial step, that direction will linger as nothing more than a fleeting hope.

Without a direction, salvation is uncertain!

The three voices and the Inner voice

There is **one action** and **three primary reactions** inside us.
There is **one force** and **three energies** inside us.
There is a **will** and **three primary emotions** inside us.
There is our **inner voice** and **three other voices** inside us.

There are **four voices** in your interior world: the voice of your inner voice and the voices of your three warriors. They are reflections of your **spark**.

The Protector says NO to anything you dislike.
The Warner says BAD to anything that might be danger.
The Savior says GOOD to anything that might be a salvation.

There is a lot of talking. If you are childish, they are childish – the interior world might suffer. You might suffer. Most people suffer more in the interior world than in the exterior world.

When four debating entities coexist within, in order to avoid inner conflict, a **leader** must emerge. A leader capable of negotiating, organizing, and offering resolutions. The title of a leader must be given to the one who can make a difference, the only one with the force that can initiate movement.

The title and role of the leader must be bestowed upon the force of the spirit – the **will**.

Through meditation, we discover our inner voice and reactions – we discover our self. Meditation allows us to initiate a dialogue within our reactions, reaching a mutual understanding. The **warriors** offer their counsel, leaving the **will** to take decisive action.

Talking in the interior world is a reflection of thinking. In order for one's interior voice to become a good talker, one must become a good listener.

> The three warriors have the sole purpose to fight darkness.
> They are all in the leadership of the will.

THE WORLD BELONGS TO THE STRONG

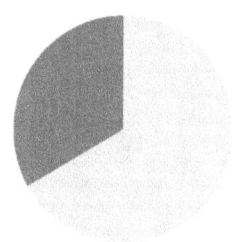

MASTERING DARKNESS

Darkness is the layer behind our world of matter, energy, and forces. It envelops and encompasses the world within it. Due to its contraction traits, darkness can influence our world.

Energy exhibits a remarkable sensitivity in detecting and responding instantaneously to darkness.

The spirit is design to be the perfect counterforce to darkness.

One cannot counter something that one does not understand. **Darkness must be understood first**.

There are 3 black dots of darkness.
- Chaos
- Loss
- Death

Chaos agitates the **reactions**.
Loss agitates the **will**.
Death agitates the **spark**.

Darkness has different intensities. A greater intensity of darkness extends further into the source of the spirit.

Darkness and energy are inherent aspects of the universe. Darkness cannot vanquish energy, and, reciprocally, energy cannot eradicate darkness. This perpetual conflict offers but one resolution: **maturity**. It is through the maturation of the spirit that we may attain mastery over darkness, and in doing so, learn to manage it effectively.

Mastering chaos

Chaos has many manifestations, and if you are not aware of them, they can deliver pure chaos into your spirit. Where nothing makes sense, and nothing can help you. Not even your warriors.

Chaos occurs every time, every moment. The outcome of it can be pleasant or unpleasant. Darkness can introduce pure chaos with its common tactic known as **surprise**. Darkness can incite **panic** which serves the role of agitating the <u>reactions</u> inside one's spirit.

Chaos can be measured. Entropy measures the amount of chaos. A higher intensity of darkness has a higher reach to the source of the spirit.

To defend ourselves from chaos, we need to understand how it circulates. First, it incites an event or a situation. It enters our senses. We detect it. It reaches the mind. It travels through our belief. And it eventually reaches our spirit.

When it reaches our spirit, the measurement of entropy is at its maximum.

The circulation of chaos follows these steps:

1. Situation or event
2. Senses
3. Mind
4. Belief
5. Spirit

The situation or event is detected by the senses. The detected information is processed by the mind. Belief forms a perceived truth statement based on the processed information. The spirit, in turn, reacts to that statement.

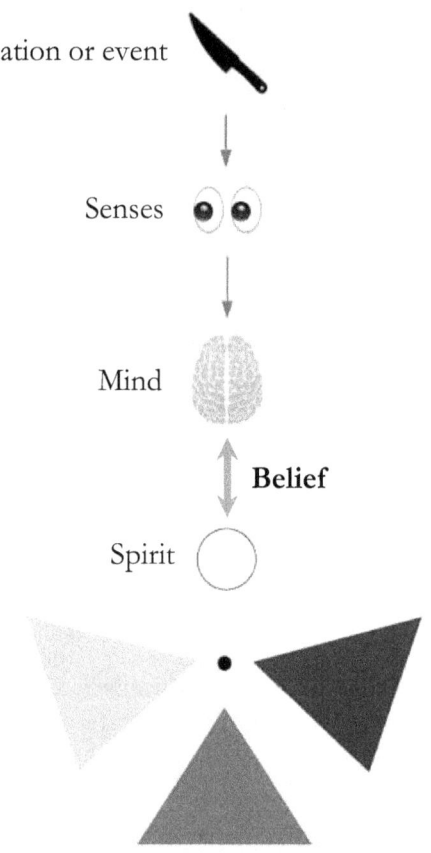

In order to counter chaos, first we need to be aware of it. We need to detect its **location** and **intensity**.

If chaos is in the **exterior world**, we can use our warriors to analyze the situation or event from which chaos manifests. When the situation or event is sufficiently observed, a conclusion can be reached.

If chaos is in the **interior world** and the entropy is high, causing the Red, Blue and Yellow reactions to lack a mature response due to their inexperience and insufficient training, there exist only one solution. The interior world needs to be disconnected from the exterior world. The bridge between the mind and the spirit needs to be disconnected. Belief needs to be disconnected.

Disconnecting the belief:
- Do not do think.
- Do not make assumptions.
- Do not believe anything.
- Just breathe and observe.

When the intensity of darkness is high, the energy in our reactions is congruently high. Reactions are a product of our own **belief**. To reduce the intensity of our energies, we need to diminish darkness from our interior world by decreasing the intensity of our belief.

As entropy decreases, we can then proceed to analyze the situation or event. Once the situation or event has been thoroughly observed, a conclusion can be drawn.

CHAOS INCITES UNBALANCE IN THE ENERGIES.

The primary tactic of chaos is **surprise**. Depending on its intensity, chaos can incite **panic**, agitating one's reactions with the explicit aim of causing an imbalance among them. When one's Warriors become unbalanced, they risk division, potentially turning them against each other. Chaos can incite conflict among one's Warriors.

The three primary reactions must remain in balance for an effective approach against darkness. The three Warriors must be reminded that they are allies of the spirit and not rivals. The Yellow Energy is not the enemy of the Blue Energy. The Red Energy is not the enemy of both. The Blue Energy is not all knowing.

One keeps its Warriors balanced by transforming one's **will** into a leader. When there is conflict inside, the will must become the **peacemaker**.

Peace is a negotiation with a resolution. A negotiation is a discussion aimed in reaching an agreement. The Yellow warrior speaks. The Blue warriors speaks. The Red warrior speaks. They are advisers of the will. Your will needs to be an attentive listener in order to form **communion** and restore balance.

There is no entity that can accurately foresee the future outcome of situations or events. Even the ones that possesses a higher form of predictability cannot guarantee certainty.

The past is gone. The present is now. **The future is unknown**.

The ones who can predict a possible future outcome have one powerful tool that reduces entropy in their spirit.

IMAGINATION

When an event or situation is observed enough, imagination can create an outcome. Imagination is the tool of creation. <u>It is the ability to create different things from the things already known</u>. Using this tool, one can create an **outcome solution** or **outcome danger**. You can also **increase** or **decrease** entropy in the interior world. It is a powerful tool, but also a dangerous one. You must know what you want. Because if you know what you want, you can break chaos.

Imagination is the most efficient counter against chaos.

Mastering loss

Loss has many manifestations, and if you are not aware of them, they can deliver pure loss into your spirit. Where suffering occurs as a result of the ongoing disruption of the connection process.

Darkness can introduce pure loss by disrupting an existing connection between one's spark with an entity. Darkness can incite **agony** which serves the role of agitating the will inside one's spirit.

Connections are stored in the spark. When you resonate with an entity that you like, you form a connection with your **will**, and it is stored in the spark. When you resonate with an entity that you dislike, you form a connection with your **red reaction**, and it is stored in the spark. When you resonate with an entity that you fear, you form a connection with your **yellow reaction**, and it is stored in the spark. When you resonate with an entity that you have faith in, you form a connection with your **blue reaction**, and it is stored in the spark.

The most powerful connections are spiritual and done with the **will**. The disrupting of those connections causes the greatest pain and decreases one's willpower.

LOSS INCITES UNBALANCE IN THE WILL.

When a connection is broken, darkness reaches and agitates the will with the goal of unbalancing it, thereby affecting the spiritual state. Loss agitates the **will** through the spark's connections.

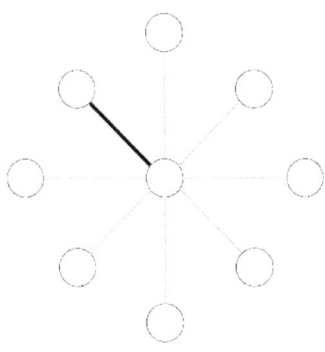

When a connection is disrupted, loss appears. When a connection is strong, loss is strong.

One might believe that abstaining from forming connections would alleviate the presence of loss. However, this is a mistake, as the willpower becomes diminished without the motivating drive provided by connections. The will is the spiritual force derived from the spark. If the spark is empty, there will be no corresponding willpower.

The appropriate response is to **value each connection**. Becoming a balanced spark.

Avoid granting undue prominence to a single connection. If this dominant connection were to disrupt, the others would lose significance, potentially leading to a descent into darkness. Do not make one soul your whole world.

Loss can be countered with a balanced spark, a balanced spiritual state and acknowledging the dark truth.

You come into this world alone. Naked – without any possessions. Except one – your vessel. That is your parents' gift to you. As time passes on your stay in this world, many other gifts will be given. Some are attained other are granted. Time gives and time takes. Gifts are temporary. They will be taken. When time comes, for you to leave this world, your vessel will be taken as well. You will leave this world as you came into it. Alone.

This is the **dark truth**. This truth provides clarity of reality to the spirit. It strengthens and empowers gratitude that you are still alive, and you have the force and energy to create significant positive change in the world, thereby increasing one's willpower.

A connection is a link between two entities. One entity is your spark. The other can be anything. When that anything disappears, that does not mean your moments, teachings and experience with that entity has disappeared as well. That does not mean that the foundational elements of the connections have been wiped from existence.

You cannot bring that anything back. But you can maintain your connection. Strengthening the values of that connection will prevent darkness from breaking it.

Strong connections are the most efficient counter against loss.

Mastering death

Death has many manifestations, and if you are not aware of them, they can deliver pure death into your spirit. Where the absence of hope devastates the entirety of one's existence.

Death is a natural part of the universe. Sooner or later, it will occur. Darkness can introduce pure death with its common tactic known as **intimidation**. Darkness can incite despair which serves the role of agitating the spark inside one's spirit.

Death has the capability to diminish the value of everything around it, rendering all else inconsequential in its presence. It is the greatest impulse given by darkness, and it has the potential to awaken the spirit's greatest power.

DEATH INCITES UNBALANCE IN THE SPARK.

When the manifestation of death infuses pure darkness into the spirit, it transcends the barriers of reactions and will, reaching and attacking the very source of the spirit.

To defend ourselves against the manifestation of death, we must strengthen the **spark**. The unbalanced spark is the one susceptible to despair when pure death is delivered.

The selfish spark values itself more. With insufficient connections, darkness can easily envelop such a solitary spark. The selfish spark is the most tormented victim.

The selfless spark values others more. With insufficient self-values encrypted in the spark, darkness can easily envelop such a vulnerable spark. The selfless spark is the easiest victim.

The balance spark values everyone. It has powerful connections and values that guards itself from the encroachment of darkness.

When the black dot death appears, one must be already ready. To defend against death, training must be done before the encounter. Many meaningful connections and values must be embedded into the spark.

The black dot **death** instigates the greatest agitation in the spirit.

It can bring chaos, infusing unbalance in the reactions. It can bring loss, infusing unbalanced in the spiritual state. It can unbalance one's spark.

The black dot death can reach the very source of the spirit.

Without a balanced spiritual state or a balanced spark to grant protection, only one defense remains.

The spirit is like a fiery water and the source of ignition of that fire is the spark. One can defend against the black dot death by mastering fire. By mastering the red energy! The Red warrior. THE PROTECTOR!

The Red Energy is the most powerful energy of the three primary energies. This is an energy that has a low frequency, and it requires a powerful will to be sustained.

Through discipline, meditation, and harnessing the power of the Red Energy, one can train to withstand threats against the spark.

Death is imminent. All the lies in the world will not release anyone from this impending event.

Nothing is immortal. Sooner or later, death reveals itself. The end of life will inevitably come, whether through accidents or natural causes. Embracing this reality rather than denying or ignoring it unveils the world's most precious currency – TIME.

Awareness of the estimated time left enhances the branches of imagination within the interior world. Depending on the maturity of one's reactions, the interior world may either suffer or flourish.

Increasing the belief that death is real, increases imagination – the ability to create. Using this ability, one can start creating the greatest purpose for the will.

Death is the greatest impulse bestowed by darkness, granting the spirit its ultimate power – the power to create!

Creation is the most efficient counter against death.

MASTERING LIGHT

Judgement

The **will** is the driving force of the spirit. It is the ability to make decisions. When faced with a situation or event, making decisions can be challenging. In such moments, our reactions offer valuable counsel.

The common encounter with darkness is chaos.

For the spirit, chaos is a **fulcrum**. Darkness does not destroy; it exerts influence upon the world, thereby instigating change. Thus, when there is a problem, there is also a solution.

The Warner and Savior are opposite balancing energies. As one energy ascends, the other declines. To enhance one, it is essential to analyze and understand the dynamics of the other.

The communication between balanced energies, aiming to guide towards a **satisfactory direction**, is referred to as JUDGEMENT.

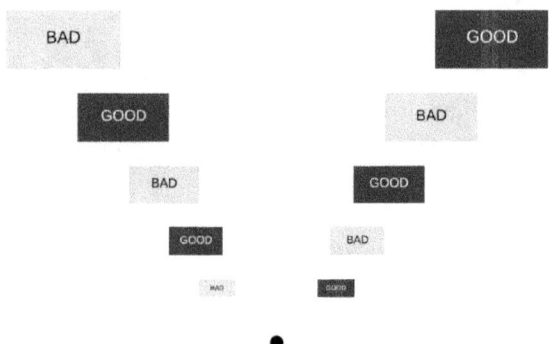

Darkness does not destroy. It changes. Change creates problems and solutions. Every problem has a solution. Every solution has a problem. The solution lies in the problem. The problem lies in the solution. There is always a solution. There is always a problem.

The Yellow reaction detects the problem. Through a comprehensive analysis of the problem, we gain a deeper understanding of its causality, and within that causality, the Blue reactions detects the solution.

If the direction offered by the Savior is a **satisfactory** one, the only remaining step is for the <u>will</u> to act.

There is no perfect solution or flawless problem. When our will analyzes a problem, the Savior eventually identifies a solution. Conversely, when our will scrutinizes the solution, the Warner eventually identifies a problem. The persistence in this overthinking approach leads to the unraveling of increasingly complex layers of chaos. Upon encountering a <u>satisfactory direction</u>, a decision must be made. Decisions serve as the disruptors of chaos. Our will, therefore, acts as the **CHAOS BREAKER**!

Two powerful methods to empower faith

FEAR and **FAITH** are balancing energies, akin to "equal siblings". They are the calibrators of truth. Fear may appear stronger at times due to a specific aspect – it possesses a "problem net" aided by chaos. Inaction often leads to the manifestation of problems. In contrast, faith lacks a "safety net", requiring the **will** to take action for a solution to materialize.

The first method is known as **courage** – an ability attributed to the Blue Warrior. <u>It is the ability to do something despite fear</u>! It is the impulse of FAITH. It is the "jump without a net" – without a certitude or a high belief. It gives the impulse needed for the will to act and for the Blue Energy to increase in intensity.

However, certainty exists! **There is always a solution**!

The second method involves a "safety net" that provides the will with a <u>desired direction</u> for the future. This is referred to as **hope**, which empowers faith.

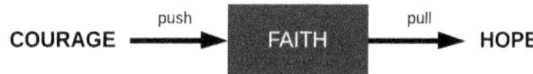

FEAR and FAITH exist to act as filters for **belief**, facilitating a deeper understanding of the exterior world and a more profound grasp of reality. When fear exhibits high amplitude, faith experiences low amplitude. Conversely, when faith demonstrates high amplitude, fear registers low amplitude. When one is low, the other is high, ensuring a constant presence of energy that can be harnessed, unless belief reaches 100% or 0%, eliminating fluctuations. Certainty serves as an impediment to the spirit, creating a blocking point. When the fulcrum lacks fluctuation, the spirit becomes immobile in the face of chaos. Certainty diminishes our capacity to adapt to change. Additionally, excessive analysis or overthinking exacerbates chaos.

A decision must be made.
The **will** is the <u>chaos breaker</u>.

If the balance is chaotic and the will is weak,
The Red warrior is our last line of defense in countering chaos.

Mastering the Red warrior

The Red Energy is the most potent among the three energies within the spirit. Its power is such that even an untrained will cannot harness or control it.

All energies originate from the **spark**, serving as reflections of it.

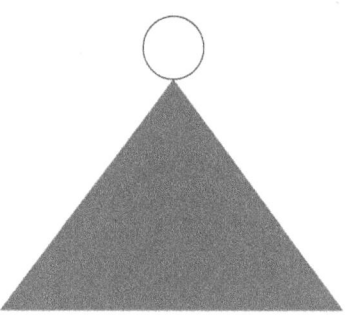

To decrease the intensity of the Red Energy, it is essential to regain control in three distinct realms.

- **Spark control**: Restoring balance in the spark to prevent the undue influence on <u>belief</u>.
- **Belief control**: Understanding the distinction between subjective and objective opinions to differentiate the <u>interior world</u> from the <u>exterior world</u>.
- **Interior world control**: Recognizing that taking things personally can have a detrimental impact on the <u>interior world</u>.

By decreasing the intensity of the Red Energy, we have accomplished a specific result.

 We have emptied the glass.

Without the energies, only the **will** remains. However, the energy provided by the will alone might not be sufficient.

So, what is to be done?

 The glass must be filled.

A reaction is an energetical response to a situation or event. To awaken the Red Energy we need to create darkness.

WARNING.

This is not a teaching.
This is a dangerous technique.

This technique unleashes the full potential of the Red Energy. One must possess knowledge and mastery over this energy to progress into the subsequent pages.

You are required to have a powerful Chaos Breaker. Your **will** stands as the sole master capable of taming and controlling the flow of the Red Energy.

A **balanced spark** is essential to prevent harm to oneself or others.

Practicing this technique will result in the introduction of darkness into your interior world.

If your Red Energy is untrained and lacks sufficient maturity, darkness will be the least of your concerns. Inviting darkness into your interior world awakens the Red Energy. It awakens the RED WARRIOR!

If the Red Energy is not mature, you **will awaken** THE RED DRAGONS, THE RED WOLVES, THE RED MONSTERS...

AND WHEN THEY ARE AWAKENED ... THEY WILL SEARCH AND DEVOUR DARKNESS...

THEY WILL DEVOUR YOUR OWN SOUL!

YOUR OWN SPIRIT WILL DEVOUR YOUR BODY AND MIND UNTIL DARKNESS IS NO MORE.

This technique is dangerous. Inviting darkness in your interior world awakens the most powerful energy in your spirit.

Darkness must be mastered in order to master LIGHT

Creating darkness

Darkness conceals itself within situations or events. We shall deliberately create a scenario, forging our **antagonist**.

The process of creation takes place within the interior world, utilizing the **interior tools**.

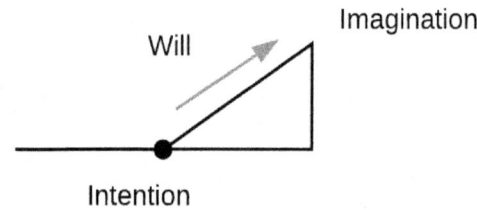

Darkness infiltrates our world through three channels: chaos, loss, and death. Among them, **DEATH** stands as the most formidable, capable of inducing both chaos and the ultimate loss – the termination of one's own life.

To create the antagonist, we need to gather all the various aspects and manifestations of death and compound them into a singular entity. We need to collect all the elements that evoke intimidation and intolerance, spanning from our earliest memories through our awakening to the present and beyond. Imagine all these things converging into one singular place, creating a powerful entity – an embodiment of our deepest challenges.

Through our **intention**, **will**, and **imagination**, we have the capacity to bring forth the supreme compound of our subjective darkness. Our subjective embodiment of death!

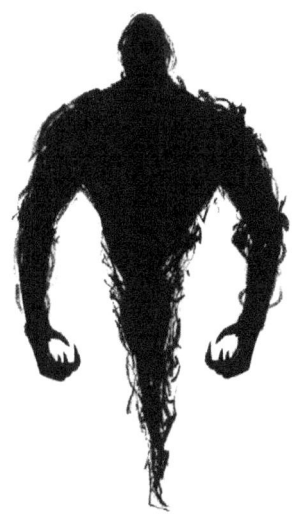

An antagonist must be deliberately created within the interior world, providing the Red warrior with a formidable adversary.

What must be remembered when creating this antagonist.
- Darkness is not a conscious entity. Your spark must not compare with it – do not mirror it.
- The Yellow warrior must not intervene. You must not fear it.
- You must not negotiate with it. You simply do not tolerate it.

After creating it, we must condense the entity. To achieve this, transform the compound darkness into a smaller, denser point.

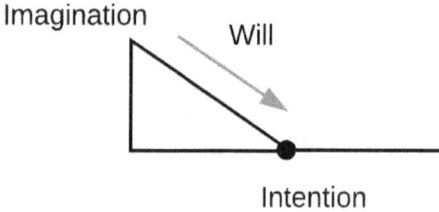

When the antagonist's influence persists, even in the **intention** phase, we have created the death dot.

Remember: You created it, and you have the power to uncreate it. You are the ruler in your interior world.

The inevitable battle

Throughout one's life development, an individual may have resolved major problems, accumulated necessary and desired resources, and achieved a form of balance akin to comfort. However, darkness remains omnipresent; it cannot be expelled, even from the tallest mountains. Sooner or later, darkness unveils itself. Chaos, loss, and death are inevitable.

One must always prepare oneself for the inevitable battle.

When the presence of the **black dot of death** exists within the interior world, the Red Energy ignites. The primary purpose of the Red warrior is to combat darkness.

The Red warrior has the ability to silence the Dark warrior, but it cannot outright defeat it, as they are considered equals.

This would be an eternal battle with no conclusive outcome.

There are two options, and one must be selected. We have the option to renounce the battle and expel the Dark warrior from our inner world, thereby laying the Red warrior to rest and calming our spirit.

There is an alternative…

The Red warrior lacks a solution to this battle. Its role is solely to serve as a protector.

To secure victory, another warrior is required. The Red warrior needs the assistance of the Blue warrior.

Two energies must undergo fusion for a more formidable warrior to emerge.

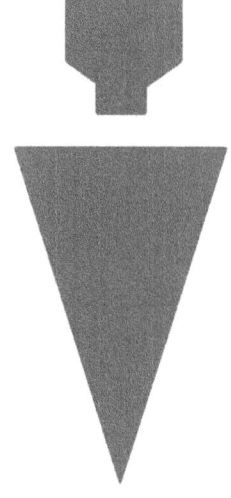

The greatest warrior within the spirit is the **Violet warrior.**

Mastering the Violet warrior

John endeavors to enhance his physical endurance for an upcoming race. Determined, he opts for a session of jogging, deciding to cover a substantial distance.

Initiating his run, John breathes deliberately, orchestrating the flow of his spirit within his body. This measured breathing prevents agitation in his spirit, allowing his body to endure the run for an extended period.

Persistently driven, John steadily paces through the distance, feeling the weight in his body accumulate. After an extensive run, signals are sent to his brain – interpreted as PAIN. Gradually, the body's resilience diminishes.

The escalating pain poses chaos for John's spirit, causing annoyance. Despite this, he maintains his resolve, breathing through the discomfort. His commitment is paramount. He has made a statement – he will finish this run. His word is important to him, and so he perseveres.

Pressing forward, John covers more than half the distance. Signals in his body intensify, accompanied by the emergence of sweat and an increase in corporal temperature.

His legs grow heavy.

Undeterred, John forges ahead. The internal struggle commences.

As John faces the challenge of exhaustion, a potential solution emerges when the **Blue warrior** suggests that he has done enough, encouraging him to stop and rest. However, the **Red Energy** flares, forcefully silencing the **Blue warrior**. John has made a statement and refuses to backtrack.

He perseveres.

Beyond the halfway point, John meditates on his commitment. With more than sufficient distance covered, the end is conceivable.

His determination prevails.

Shoulders and arms laden with fatigue, John's will struggles to sustain his body. He craves more energy.

In a decisive move, he closes his eyes, invoking the Dark warrior.

The Dark warrior speaks.
- **You are worthless!**
- **You are nothing!**
- **All your decisions have led you to this moment, and YOU WILL FAIL**

The **Red warrior** ignites, engaged in a fierce battle against the Dark warrior.

John opens his eyes, gaining newfound energy, propelling him to run even faster. However, the heightened energy risks further damage to his body. The Dark warrior needs to be expelled from the interior word.

Utilizing the **Blue warrior**, John crafts a purpose: **The Dark warrior can be defeated <u>if</u> John finishes the race**. A solution surfaces.

John firmly believes in this statement and a surge of energy streams through him, rendering the pain inconsequential. His path is clear, and his will is unwavering.

With **WILL**, **ANGER**, and **FAITH** in unison, John triumphantly completes the run.

The strength of the Violet warrior derives from the Red warrior, while the key to victory remains concealed in the mastery of the Blue warrior.

To master the Blue warrior, one requires something that seeks solutions.

THE WORLD BELONGS TO THE INTELLIGENT

Intelligence

The spirit is the energy source of every living being. It is a subconscious battery. This energy traverses through the entire body and mind granting **abilities** such as movement and thinking. Every existing ability requires effort. That energy providing the effort is from the spirit. Enhanced abilities are a result of heightened or precise effort from the spirit.

The body houses a remarkable organ, the **brain**, through which the spirit can access and utilize its abilities. The most powerful among them is intelligence.

Intelligence is a composite ability, formed from three other abilities. The three other abilities are:
- Curiosity
- Understanding
- Imagination

Our brain is a composition of neurons and axons. Reflected as **points** and **lines**. Those points and lines create a map. The map of the mind is known as knowledge. For a wondering spirit to decipher, navigate, and expand this realm, it is essential to master the abilities of the mind.

Curiosity

Curiosity is the ability to extract information.

Points are answers. Complex answers are structural. Lines that reach and hold those points together are questions.

Questions lead to answers.

When a question is put, it reaches a different **point**. From that different point, the question can travel further.

Until data becomes information.

The purpose of curiosity is to form knowledge.

Understanding

Understanding is the ability to perceive the structure of the connections.

Asking a question leads from one point to the other. Connecting many points forms a structure.

The purpose of understanding is to form knowledge.

Imagination

Imagination is the ability to create different things from the things already known.

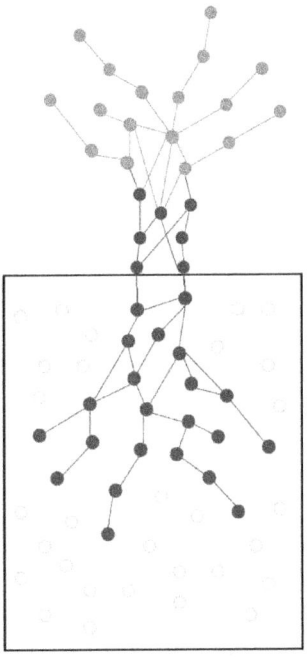

The mind functions as a data processor and a tool for creation. The creative process commences within the confines of the interior world.

The purpose of imagination is to form knowledge.

Wisdom

The world is continually changing, and our senses are in constant monitoring mode, perceiving the data. Our intelligent abilities play a crucial role in strengthening and expanding the grid map of our mind.

For our mind to effectively handle such a significant amount of processing power, it requires **wisdom**.

Let's offer an example.

$$\{\frac{8!}{10} + \frac{1}{\sqrt{25}} * [(20 * 20)^2 + 990] + \frac{5!}{24} * \sum_{k=1}^{10} 2^{3+1} * 3^k\} * 0$$

In order to solve this equation, we need intelligence.

Insight is the ability to see an event from one point to the other. To see its starting point and its ending point.

Seeing the whole equation, we can observe that at the ending point there is a multiplication operation with the value zero.

That means no matter the result inside the brackets, the final result will be zero.

After seeing the situation or event, **wisdom** <u>is intelligence and ignorance combined</u>.

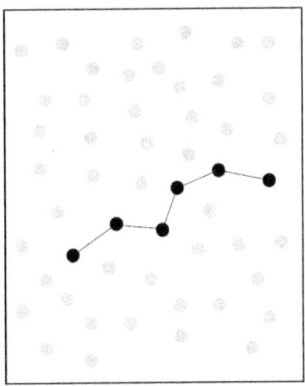

Remember: You need intelligence in order to know what to ignore.

The abilities of the mind facilitate the analytical deconstruction of a given situation or event. This process aims to comprehend its underlying purpose, the origins of its creation, and potential future ramifications. The predictive capabilities derived from such analysis offer insight into forthcoming situations or events.

The inherent limitation of intelligence lies in its inability to reverse alterations in the external world. A supplementary element is required to address this constraint.

THE WORLD BELONGS TO THE ADAPTOR

The three types of situations & events

The world does not belong neither to the strong, neither to the intelligent. It belongs to the **adaptor**. But one must be strong and intelligent to become an adaptor.

The first law of the world is **that everything changes.**

In the interior world, our control is limitless.
In the exterior world, our control is limited.

From the beginning to the end of one's life, countless situations and events will be encountered.

There are three types of situations & events.
- Controllable
- Influential
- Uncontrollable

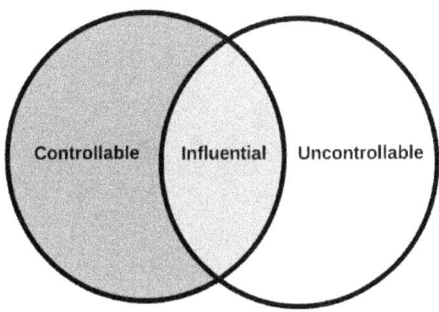

Life is a composition of moments.

Each moment unfolds a unique situation or event. Identifying the type empowers informed decision-making.

For situations under your <u>control</u> or <u>influence</u>, inquire your will:
– **What do I want from this moment?**

For those beyond control, acceptance is the key.

When your belief system hovers at 99.9%, adaptation to change flows effortlessly. However, at 100%, it transforms into certainty, creating a blocking point. When change appears, breaking and reconstructing this point with updated understanding becomes imperative.

Maintaining a volatile point at 99.9% might be challenging, yet it swiftly adapts to change.

THE APOGEE OF THE SPIRIT

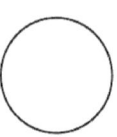

PROBLEM SOLVER

The universe is not just filled with sunshine, rainbows, and lollipops; it also encompasses darkness. The purpose of the spirit is to evolve, maturing its **will** and **warriors**, thereby acquiring a formidable defense against darkness.

A mature spirit can effectively counter the appearance of darkness.

An untrained and immature spirit exhibits the behavior of a child. The inherent purpose of a child is to mature. Failing to do so leaves the spirit vulnerable to the merciless grasp of darkness, which will inevitably devour it.

Every component associated with the spirit must mature for the entire spirit to attain its greatest form.

Maturity in belief transforms you into an **ADAPTOR**, rendering you open and receptive to change. Mature reactions shape you as a **WARRIOR**, with the primary goal of ending conflicts. A mature will elevates you to a **LEADER**, perhaps not in terms of leading others, but certainly as a leader of oneself.

Mastering darkness grants enhanced abilities as well. Maturity in chaos makes you a **VISIONARY**. Maturity in loss makes you a **PEACEMAKER**. Maturity in death makes you a **CREATOR**.

Maturing every component embodies all the characteristics of a PROBLEM SOLVER, encompassing analytical thinking, adaptability, and the capability to devise innovative solutions in the face of challenges.

BALANCE IS THE WAY

The purpose of energy

The universe has a way of humbling those who are excessively selfish and can lead to feelings of resentment for those who are overly selfless. Because we are intricately connected with the universe. Electrons are the hidden life force of the universe. They are omnipresent and can be found in every corner of our existence. Electrons are the hidden life force of ourselves. They travel inside the very atoms of our body and form our spirit. Through quantum phenomena, electrons have the ability to become entangled as a result of their interaction with electromagnetic forces. Thus, there exists a profound interconnectedness among all things, and they are composed of energy. The purpose of energy is to achieve stability, allowing it to endure for an extended duration. To maintain this stability, the universe responds to any actions and reactions that are imbalanced by readjusting itself.

The manifestations of balance encompass numerous representations, including <u>equilibrium</u>, <u>equality</u>, <u>equity</u>, <u>fairness</u>, <u>justness</u>, <u>justice</u>, <u>stability</u>, and the most known of them all, **liberty**.

The purpose of energy is to attain balance. Every spirit desires it, even those with malevolent intentions seek what is fair.

HOME

The gift from our parents

From the day we are born until we embrace our last day on this world, we own only one true possession. That possession is received as a gift from our parents, and it is our own physical **body**.

Our body is the home of the spirit. It is the only real possession it truly has. Besides that, everything else are temporary resources.

One way to honor this gift is by respecting and cherishing it. A well-cared home is a reflection of a self-respected resident. Self-care is self-respect.

The most crucial aspect of our existence is our spirit, as it provides us with life and vitality. Our vessels offer a lasting abode for our spirit in this world. To ensure a qualitatively enduring life, we need to fortify our values.

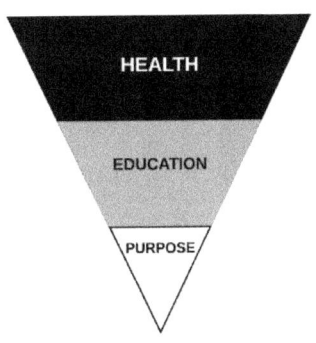

WE ARE NOT ALONE

UNITY

This world is truly mysterious. It is filled with wonders and diverse creatures, each with its unique characteristics and traits.

There is so much diversity.

But there is something that connects us.

All living beings have something in common. All **humans**, **animals**, **plants**, as different as they are, there is something in which we can find a resemblance.

There is something inside each one of us that somehow unites us.

We each have a **spirit** inside.

A spirit that possesses an innate awareness, unburdened by education or complexity, in identifying its true adversary. Surprisingly, that foe is not its siblings.

Darkness has a wicked way to divide us, to create conflict and hatred between us.

Stay strong. We are brothers and sisters.

We are the siblings.

www.ingramcontent.com/pod-product-compliance
Lightning Source LLC
Chambersburg PA
CBHW070646220526
45466CB00001B/322